Eureka Math™

Grade 1 Fluency
Modules 4–6

Published by Great Minds®.

Copyright © 2015 Great Minds®. No part of this work may be reproduced, sold, or commercialized, in whole or in part, without written permission from Great Minds®. Noncommercial use is licensed pursuant to a Creative Commons Attribution-NonCommercial-ShareAlike 4.0 license; for more information, go to http://greatminds.org/copyright. *Great Minds* and *Eureka Math* are registered trademarks of Great Minds®.

Printed in the U.S.A.

This book may be purchased from the publisher at eureka-math.org.

10 9 8 7 6 5

ISBN 978-1-64054-621-9

G1-M4-M6-P/F-04.2018

Learn ◆ Practice ◆ Succeed

Eureka Math™ student materials for *A Story of Units*® (K–5) are available in the *Learn, Practice, Succeed* trio. This series supports differentiation and remediation while keeping student materials organized and accessible. Educators will find that the *Learn, Practice, and Succeed* series also offers coherent—and therefore, more effective—resources for Response to Intervention (RTI), extra practice, and summer learning.

Learn

Eureka Math Learn serves as a student's in-class companion where they show their thinking, share what they know, and watch their knowledge build every day. *Learn* assembles the daily classwork—Application Problems, Exit Tickets, Problem Sets, templates—in an easily stored and navigated volume.

Practice

Each *Eureka Math* lesson begins with a series of energetic, joyous fluency activities, including those found in *Eureka Math Practice*. Students who are fluent in their math facts can master more material more deeply. With *Practice*, students build competence in newly acquired skills and reinforce previous learning in preparation for the next lesson.

Together, *Learn* and *Practice* provide all the print materials students will use for their core math instruction.

Succeed

Eureka Math Succeed enables students to work individually toward mastery. These additional problem sets align lesson by lesson with classroom instruction, making them ideal for use as homework or extra practice. Each problem set is accompanied by a Homework Helper, a set of worked examples that illustrate how to solve similar problems.

Teachers and tutors can use *Succeed* books from prior grade levels as curriculum-consistent tools for filling gaps in foundational knowledge. Students will thrive and progress more quickly as familiar models facilitate connections to their current grade-level content.

Students, families, and educators:

Thank you for being part of the *Eureka Math*™ community, where we celebrate the joy, wonder, and thrill of mathematics. One of the most obvious ways we display our excitement is through the fluency activities provided in *Eureka Math Practice*.

What is fluency in mathematics?

You may think of *fluency* as associated with the language arts, where it refers to speaking and writing with ease. In prekindergarten through grade 5, the *Eureka Math* curriculum contains multiple daily opportunities to build fluency *in mathematics*. Each is designed with the same notion—growing every student's ability to use mathematics *with ease*. Fluency experiences are generally fast-paced and energetic, celebrating improvement and focusing on recognizing patterns and connections within the material. They are not intended to be graded.

Eureka Math fluency activities provide differentiated practice through a variety of formats—some are conducted orally, some use manipulatives, others use a personal whiteboard, and still others use a handout and paper-and-pencil format. *Eureka Math Practice* provides each student with the printed fluency exercises for his or her grade level.

What is a Sprint?

Many printed fluency activities utilize the format we call a Sprint. These exercises build speed and accuracy with already acquired skills. Used when students are nearing optimum proficiency, Sprints leverage tempo to build a low-stakes adrenaline boost that increases memory and recall. Their intentional design makes Sprints inherently differentiated; the problems build from simple to complex, with the first quadrant of problems being the simplest and each subsequent quadrant adding complexity. Further, intentional patterns within the sequence of problems engage students' higher order thinking skills.

The suggested format for delivering a Sprint calls for students to do two consecutive Sprints (labeled A and B) on the same skill, each timed at one minute. Students pause between Sprints to articulate the patterns they noticed as they worked the first Sprint. Noticing the patterns often provides a natural boost to their performance on the second Sprint.

Sprints can be conducted with an untimed protocol as well. The untimed protocol is highly recommended when students are still building confidence with the level of complexity of the first quadrant of problems. Once all students are prepared for success on the Sprint, the work of improving speed and accuracy with the energy of a timed protocol is often welcome and invigorating.

Where can I find other fluency activities?

The *Eureka Math Teacher Edition* guides educators in the delivery of all fluency activities for each lesson, including those that do not require print materials. Additionally, the *Eureka Digital Suite* provides access to the fluency activities for all grade levels, searchable by standard or lesson.

Best wishes for a year filled with aha moments!

Jill Diniz

Jill Diniz
Director of Mathematics
Great Minds

Contents

Module 4

Module 5

Module 6

Grade 1
Module 4

A

Name _____ Date _____

Number Correct: [ ]

*Write the missing number. Pay attention to the addition or subtraction sign.

1	$5 + 1 = \square$		16	$29 + 10 = \square$	
2	$15 + 1 = \square$		17	$9 + 1 = \square$	
3	$25 + 1 = \square$		18	$19 + 1 = \square$	
4	$5 + 10 = \square$		19	$29 + 1 = \square$	
5	$15 + 10 = \square$		20	$39 + 1 = \square$	
6	$25 + 10 = \square$		21	$40 - 1 = \square$	
7	$8 - 1 = \square$		22	$30 - 1 = \square$	
8	$18 - 1 = \square$		23	$20 - 1 = \square$	
9	$28 - 1 = \square$		24	$20 + \square = 21$	
10	$38 - 1 = \square$		25	$20 + \square = 30$	
11	$38 - 10 = \square$		26	$27 + \square = 37$	
12	$28 - 10 = \square$		27	$27 + \square = 28$	
13	$18 - 10 = \square$		28	$\square + 10 = 34$	
14	$9 + 10 = \square$		29	$\square - 10 = 14$	
15	$19 + 10 = \square$		30	$\square - 10 = 24$	

B

Number Correct:

Name _____   Date _____

*Write the missing number. Pay attention to the addition or subtraction sign.

1	$4 + 1 = \square$		16	$28 + 10 = \square$	
2	$14 + 1 = \square$		17	$9 + 1 = \square$	
3	$24 + 1 = \square$		18	$19 + 1 = \square$	
4	$6 + 10 = \square$		19	$29 + 1 = \square$	
5	$16 + 10 = \square$		20	$39 + 1 = \square$	
6	$26 + 10 = \square$		21	$40 - 1 = \square$	
7	$7 - 1 = \square$		22	$30 - 1 = \square$	
8	$17 - 1 = \square$		23	$20 - 1 = \square$	
9	$27 - 1 = \square$		24	$10 + \square = 11$	
10	$37 - 1 = \square$		25	$10 + \square = 20$	
11	$37 - 10 = \square$		26	$22 + \square = 32$	
12	$27 - 10 = \square$		27	$22 + \square = 23$	
13	$17 - 10 = \square$		28	$\square + 10 = 39$	
14	$8 + 10 = \square$		29	$\square - 10 = 19$	
15	$18 + 10 = \square$		30	$\square - 10 = 29$	

EUREKA MATH™

Lesson 7:   Compare two quantities, and identify the greater or lesser of the two given numerals.

13

©2015 Great Minds®. eureka-math.org

tens	ones

large place value chart

Lesson 7:   Compare two quantities, and identify the greater or lesser of the two given numerals.

15

©2015 Great Minds®. eureka-math.org

Name _____   Date _____

Core Subtraction Fluency Review

1. $8 - 0 =$ ____
2. $8 - 1 =$ ____
3. $7 - 7 =$ ____
4. $3 - 3 =$ ____
5. $3 - 2 =$ ____
6. $4 - 2 =$ ____
7. $5 - 2 =$ ____
8. $5 - 3 =$ ____
9. $9 - 2 =$ ____
10. $8 - 2 =$ ____
11. $7 - 2 =$ ____
12. $4 - 4 =$ ____
13. $4 - 3 =$ ____
14. $5 - 4 =$ ____
15. $8 - 3 =$ ____

16. $9 - 3 =$ ____
17. $10 - 3 =$ ____
18. $10 - 4 =$ ____
19. $10 - 2 =$ ____
20. $10 - 8 =$ ____
21. $10 - 7 =$ ____
22. $10 - 6 =$ ____
23. $6 - 6 =$ ____
24. $7 - 7 =$ ____
25. $7 - 6 =$ ____
26. $8 - 8 =$ ____
27. $8 - 7 =$ ____
28. $9 - 9 =$ ____
29. $9 - 8 =$ ____
30. $10 - 9 =$ ____

31. $5 - 5 =$ ____
32. $6 - 5 =$ ____
33. $7 - 5 =$ ____
34. $8 - 5 =$ ____
35. $8 - 4 =$ ____
36. $10 - 5 =$ ____
37. $9 - 5 =$ ____
38. $9 - 4 =$ ____
39. $6 - 3 =$ ____
40. $6 - 4 =$ ____
41. $7 - 3 =$ ____
42. $7 - 4 =$ ____
43. $8 - 6 =$ ____
44. $9 - 6 =$ ____
45. $9 - 7 =$ ____

A

Name _____   Date _____

Number Correct: ⬳

*Write the missing number in the sequence.

1.	0, 1, 2, ___		16.	15, ___, 13, 12	
2.	10, 11, 12, ___		17.	___, 24, 23, 22	
3.	20, 21, 22, ___		18.	6, 16, ___, 36	
4	10, 9, 8, ___		19.	7, ___, 27, 37	
5	20, 19, 18, ___		20.	___, 19, 29, 39	
6.	40, 39, 38, ___		21.	___, 26, 16, 6	
7.	0, 10, 20, ___		22.	34, ___, 14, 4	
8.	2, 12, 22, ___		23.	___, 20, 21, 22	
9.	5, 15, 25, ___		24.	29, ___, 31, 32	
10.	40, 30, 20, ___		25.	5, ___, 25, 35	
11.	39, 29, 19, ___		26.	___, 25, 15, 5	
12.	7, 8, 9, ___		27.	2, 4, ___, 8	
13.	7, 8, ___, 10		28.	___, 14, 16, 18	
14.	17, ___, 19, 20		29.	8, ___, 4, 2	
15.	15, 14, ___, 12		30.	___, 18, 16, 14	

EUREKA MATH™

B

Number Correct:

Name _____   Date _____

*Write the missing number in the sequence.

1.	1, 2, 3, __		16.	13, __, 11, 10	
2.	11, 12, 13, __		17.	__, 22, 21, 20	
3.	21, 22, 23, __		18.	5, 15, __, 35	
4.	10, 9, 8, __		19.	4, __, 24, 34	
5.	20, 19, 18, __		20.	__, 17, 27, 37	
6.	30, 29, 28, __		21.	__, 29, 19, 9	
7.	0, 10, 20, __		22.	31, __, 11, 1	
8.	3, 13, 23, __		23.	__, 30, 31, 32	
9.	6, 16, 26, __		24.	19, __, 21, 22	
10.	40, 30, 20, __		25.	5, __, 25, 35	
11.	38, 28, 18, __		26.	__, 25, 15, 5	
12.	6, 7, 8, __		27.	2, 4, __, 8	
13.	6, 7, __, 9		28.	__, 12, 14, 16	
14.	16, __, 18, 19		29.	12, __, 8, 6	
15.	16, __, 14, 13		30.	__, 20, 18, 16	

EUREKA MATH

Lesson 10:   Use the symbols >, =, and < to compare quantities and numerals.

21

A

Number Correct: ⬡

Name _____   Date _____

*Write the missing number. Pay attention to the + and – signs.

1.	$3 + \square = 4$		16.	$3 + \square = 7$	
2.	$1 + \square = 4$		17.	$7 = 4 + \square$	
3.	$4 - 1 = \square$		18.	$7 - 4 = \square$	
4.	$4 - 3 = \square$		19.	$7 - 3 = \square$	
5.	$3 + \square = 5$		20.	$3 + \square = 8$	
6.	$2 + \square = 5$		21.	$8 = 5 + \square$	
7.	$5 - 2 = \square$		22.	$\square = 8 - 5$	
8.	$5 - 3 = \square$		23.	$\square = 8 - 3$	
9.	$4 + \square = 6$		24.	$3 + \square = 9$	
10.	$2 + \square = 6$		25.	$9 = 6 + \square$	
11.	$6 - 2 = \square$		26.	$\square = 9 - 6$	
12.	$6 - 4 = \square$		27.	$\square = 9 - 3$	
13.	$6 - 3 = \square$		28.	$9 - 4 = \square + 2$	
14.	$3 + \square = 6$		29.	$\square + 3 = 9 - 3$	
15.	$6 - \square = 3$		30.	$\square - 7 = 8 - 6$	

B

Number Correct: _____

Name _____   Date _____

*Write the missing number. Pay attention to the + and – signs.

1.	$4 + \square = 4$		16.	$2 + \square = 7$	
2.	$0 + \square = 4$		17.	$7 = 5 + \square$	
3.	$4 - 0 = \square$		18.	$7 - 5 = \square$	
4.	$4 - 4 = \square$		19.	$7 - 2 = \square$	
5.	$4 + \square = 5$		20.	$2 + \square = 8$	
6.	$1 + \square = 5$		21.	$8 = 6 + \square$	
7.	$5 - 1 = \square$		22.	$\square = 8 - 6$	
8.	$5 - 4 = \square$		23.	$\square = 8 - 2$	
9.	$5 + \square = 6$		24.	$2 + \square = 9$	
10.	$1 + \square = 6$		25.	$9 = 7 + \square$	
11.	$6 - 1 = \square$		26.	$\square = 9 - 7$	
12.	$6 - 5 = \square$		27.	$\square = 9 - 2$	
13.	$2 + \square = 6$		28.	$9 - 3 = \square + 3$	
14.	$4 + \square = 6$		29.	$\square + 2 = 9 - 4$	
15.	$6 - 4 = \square$		30.	$\square - 6 = 8 - 3$	

Name _____   Date _____

Core Addition Fluency Review: Missing Addends

1. $5 + \underline{\hspace{1cm}} = 5$

2. $4 + \underline{\hspace{1cm}} = 5$

3. $2 + \underline{\hspace{1cm}} = 5$

4. $3 + \underline{\hspace{1cm}} = 5$

5. $0 + \underline{\hspace{1cm}} = 5$

6. $1 + \underline{\hspace{1cm}} = 5$

7. $1 + \underline{\hspace{1cm}} = 6$

8. $0 + \underline{\hspace{1cm}} = 6$

9. $6 + \underline{\hspace{1cm}} = 6$

10. $5 + \underline{\hspace{1cm}} = 6$

11. $3 + \underline{\hspace{1cm}} = 6$

12. $4 + \underline{\hspace{1cm}} = 6$

13. $2 + \underline{\hspace{1cm}} = 6$

14. $2 + \underline{\hspace{1cm}} = 7$

15. $5 + \underline{\hspace{1cm}} = 7$

16. $6 + \underline{\hspace{1cm}} = 7$

17. $1 + \underline{\hspace{1cm}} = 7$

18. $0 + \underline{\hspace{1cm}} = 7$

19. $7 + \underline{\hspace{1cm}} = 7$

20. $3 + \underline{\hspace{1cm}} = 7$

21. $4 + \underline{\hspace{1cm}} = 7$

22. $4 + \underline{\hspace{1cm}} = 8$

23. $5 + \underline{\hspace{1cm}} = 8$

24. $6 + \underline{\hspace{1cm}} = 8$

25. $2 + \underline{\hspace{1cm}} = 8$

26. $3 + \underline{\hspace{1cm}} = 8$

27. $0 + \underline{\hspace{1cm}} = 8$

28. $8 + \underline{\hspace{1cm}} = 8$

29. $7 + \underline{\hspace{1cm}} = 8$

30. $1 + \underline{\hspace{1cm}} = 8$

31. $9 + \underline{\hspace{1cm}} = 9$

32. $0 + \underline{\hspace{1cm}} = 9$

33. $1 + \underline{\hspace{1cm}} = 9$

34. $2 + \underline{\hspace{1cm}} = 9$

35. $7 + \underline{\hspace{1cm}} = 9$

36. $6 + \underline{\hspace{1cm}} = 9$

37. $5 + \underline{\hspace{1cm}} = 9$

38. $3 + \underline{\hspace{1cm}} = 9$

39. $4 + \underline{\hspace{1cm}} = 9$

40. $4 + \underline{\hspace{1cm}} = 10$

41. $5 + \underline{\hspace{1cm}} = 10$

42. $6 + \underline{\hspace{1cm}} = 10$

43. $3 + \underline{\hspace{1cm}} = 10$

44. $1 + \underline{\hspace{1cm}} = 10$

45. $2 + \underline{\hspace{1cm}} = 10$

A

Number Correct:

Name _____   Date _____

*Write the missing number.

1	6 + 1 = ☐		16	6 + 3 = ☐	
2	16 + 1 = ☐		17	16 + 3 = ☐	
3	26 + 1 = ☐		18	26 + 3 = ☐	
4	5 + 2 = ☐		19	4 + 5 = ☐	
5	15 + 2 = ☐		20	15 + 4 = ☐	
6	25 + 2 = ☐		21	8 + 2 = ☐	
7	5 + 3 = ☐		22	18 + 2 = ☐	
8	15 + 3 = ☐		23	28 + 2 = ☐	
9	25 + 3 = ☐		24	8 + 3 = ☐	
10	4 + 4 = ☐		25	8 + 13 = ☐	
11	14 + 4 = ☐		26	8 + 23 = ☐	
12	24 + 4 = ☐		27	8 + 5 = ☐	
13	5 + 4 = ☐		28	8 + 15 = ☐	
14	15 + 4 = ☐		29	28 + ☐ = 33	
15	25 + 4 = ☐		30	25 + ☐ = 33	

EUREKA MATH™

Lesson 19: Use tape diagrams as representations to solve *put together/take apart with total unknown* and *add to with result unknown* word problems.

29

B

Number Correct: ⭐

Name _____ Date _____

*Write the missing number.

1	5 + 1 = ☐		16	6 + 3 = ☐	
2	15 + 1 = ☐		17	16 + 3 = ☐	
3	25 + 1 = ☐		18	26 + 3 = ☐	
4	4 + 2 = ☐		19	3 + 5 = ☐	
5	14 + 2 = ☐		20	15 + 3 = ☐	
6	24 + 2 = ☐		21	9 + 1 = ☐	
7	5 + 3 = ☐		22	19 + 1 = ☐	
8	15 + 3 = ☐		23	29 + 1 = ☐	
9	25 + 3 = ☐		24	9 + 2 = ☐	
10	6 + 2 = ☐		25	9 + 12 = ☐	
11	16 + 2 = ☐		26	9 + 22 = ☐	
12	26 + 2 = ☐		27	9 + 5 = ☐	
13	4 + 3 = ☐		28	9 + 15 = ☐	
14	14 + 3 = ☐		29	29 + ☐ = 34	
15	24 + 3 = ☐		30	25 + ☐ = 34	

EUREKA MATH

Lesson 19: Use tape diagrams as representations to solve *put together/take apart with total unknown* and *add to with result unknown* word problems.

31

A

Number Correct: _____

Name _____   Date _____

*Write the missing number. Pay attention to the + and – signs.

1	2 + 2 = ☐		16	2 + ☐ = 8	
2	2 + ☐ = 4		17	6 + ☐ = 8	
3	4 - 2 = ☐		18	8 - 6 = ☐	
4	3 + 3 = ☐		19	8 - 2 = ☐	
5	3 + ☐ = 6		20	9 + 2 = ☐	
6	6 - 3 = ☐		21	9 + ☐ = 11	
7	4 + ☐ = 7		22	11 - 9 = ☐	
8	3 + ☐ = 7		23	9 + ☐ = 15	
9	7 - 3 = ☐		24	15 - 9 = ☐	
10	7 - 4 = ☐		25	8 + ☐ = 15	
11	5 + 4 = ☐		26	15 - ☐ = 8	
12	4 + ☐ = 9		27	8 + ☐ = 17	
13	9 - 4 = ☐		28	17 - ☐ = 8	
14	9 - 5 = ☐		29	27 - ☐ = 8	
15	9 - ☐ = 4		30	37 - ☐ = 8	

B

Number Correct:

Name _____   Date _____

*Write the missing number. Pay attention to the + and – signs.

1	3 + 3 = □		16	2 + □ = 9	
2	3 + □ = 6		17	7 + □ = 9	
3	6 - 3 = □		18	9 - 7 = □	
4	4 + 4 = □		19	9 - 2 = □	
5	4 + □ = 8		20	9 + 5 = □	
6	8 - 4 = □		21	9 + □ = 14	
7	4 + □ = 9		22	14 - 9 = □	
8	5 + □ = 9		23	9 + □ = 16	
9	9 - 5 = □		24	16 - 9 = □	
10	9 - 4 = □		25	8 + □ = 16	
11	3 + 4 = □		26	16 - □ = 8	
12	4 + □ = 7		27	8 + □ = 16	
13	7 - 4 = □		28	16 - □ = 8	
14	7 - 3 = □		29	26 - □ = 8	
15	7 - □ = 3		30	36 - □ = 8	

My Addition Practice

1. 6 + 0 = ___	11. 7 + 1 = ___	21. 5 + 3 = ___
2. 0 + 6 = ___	12. ___ = 1 + 7	22. ___ = 5 + 4
3. 5 + 1 = ___	13. 3 + 3 = ___	23. 6 + 4 = ___
4. 1 + 5 = ___	14. 3 + 4 = ___	24. 4 + 6 = ___
5. 6 + 1 = ___	15. ___ = 3 + 5	25. ___ = 4 + 4
6. 1 + 6 = ___	16. 6 + 3 = ___	26. 3 + 4 = ___
7. 6 + 2 = ___	17. 7 + 3 = ___	27. 5 + 5 = ___
8. 5 + 2 = ___	18. ___ = 7 + 2	28. ___ = 4 + 5
9. 2 + 5 = ___	19. 2 + 7 = ___	29. 3 + 7 = ___
10. 2 + 4 = ___	20. 2 + 8 = ___	30. ___ = 3 + 6

Today, I finished _____ problems.

I solved _____ problems correctly.

Lesson 23:   Interpret two-digit numbers as tens and ones, including cases with more than 9 ones.

37

Name _____   Date _____

My Missing Addend Practice

1. $6 + \underline{\hspace{1cm}} = 6$

2. $0 + \underline{\hspace{1cm}} = 6$

3. $5 + \underline{\hspace{1cm}} = 6$

4. $4 + \underline{\hspace{1cm}} = 6$

5. $0 + \underline{\hspace{1cm}} = 7$

6. $6 + \underline{\hspace{1cm}} = 7$

7. $1 + \underline{\hspace{1cm}} = 7$

8. $7 + \underline{\hspace{1cm}} = 8$

9. $1 + \underline{\hspace{1cm}} = 8$

10. $6 + \underline{\hspace{1cm}} = 8$

11. $3 + \underline{\hspace{1cm}} = 6$

12. $4 + \underline{\hspace{1cm}} = 8$

13. $10 = 5 + \underline{\hspace{1cm}}$

14. $5 + \underline{\hspace{1cm}} = 9$

15. $5 + \underline{\hspace{1cm}} = 7$

16. $8 = 5 + \underline{\hspace{1cm}}$

17. $5 + \underline{\hspace{1cm}} = 9$

18. $8 + \underline{\hspace{1cm}} = 10$

19. $7 + \underline{\hspace{1cm}} = 10$

20. $10 = 6 + \underline{\hspace{1cm}}$

21. $4 + \underline{\hspace{1cm}} = 7$

22. $7 = 3 + \underline{\hspace{1cm}}$

23. $2 + \underline{\hspace{1cm}} = 7$

24. $2 + \underline{\hspace{1cm}} = 8$

25. $9 = 2 + \underline{\hspace{1cm}}$

26. $2 + \underline{\hspace{1cm}} = 10$

27. $10 = 3 + \underline{\hspace{1cm}}$

28. $3 + \underline{\hspace{1cm}} = 9$

29. $4 + \underline{\hspace{1cm}} = 9$

30. $10 = 4 + \underline{\hspace{1cm}}$

Today, I finished _____ problems.

I solved _____ problems correctly.

Lesson 23:   Interpret two-digit numbers as tens and ones, including cases with
more than 9 ones.

39

Name _____   Date _____

My Related Addition and Subtraction Practice

1. 5 + ___ = 6

2. 1 + ___ = 6

3. 6 – 1 = ___

4. 9 + ___ = 10

5. 1 + ___ = 10

6. 10 – 9 = ___

7. 5 + ___ = 10

8. 10 – 5 = ___

9. 8 + ___ = 10

10. 10 – 8 = ___

11. 7 + ___ = 10

12. 10 – 7 = ___

13. 5 + ___ = 7

14. 7 – 5 = ___

15. 5 + ___ = 8

16. 8 – 5 = ___

17. 4 + ___ = 6

18. 6 – 4 = ___

19. 3 + ___ = 6

20. 6 – 3 = ___

21. 4 + ___ = 8

22. 8 – 4 = ___

23. 4 + ___ = 7

24. 7 – 4 = ___

25. 5 + ___ = 9

26. 9 – 5 = ___

27. 6 + ___ = 9

28. 9 – 6 = ___

29. 4 + ___ = 7

30. 7 – 4 = ___

Today, I finished _____ problems.

I solved _____ problems correctly.

Lesson 23:   Interpret two-digit numbers as tens and ones, including cases with more than 9 ones.

41

©2015 Great Minds®. eureka-math.org

Name _____   Date _____

My Subtraction Practice

1. 6 - 0 = ____	11. 6 - 3 = ____	21. 8 - 4 = ____
2. 6 - 1 = ____	12. 7 - 3 = ____	22. 8 - 3 = ____
3. 7 - 1 = ____	13. 9 – 3 = ____	23. 8 - 5 = ____
4. 8 - 1 = ____	14. 10 - 8 = ____	24. 9 - 5 = ____
5. 6 - 2 = ____	15. 10 - 6 = ____	25. 9 - 4 = ____
6. 7 - 2 = ____	16. 10 – 4 = ____	26. 7 - 3 = ____
7. 9 - 2 = ____	17. 10 - 5 = ____	27. 10 - 7 = ____
8. 10 - 10 = ____	18. 7 – 6 = ____	28. 9 - 7 = ____
9. 10 - 9 = ____	19. 7 - 5 = ____	29. 9 - 6 = ____
10. 10 - 7 = ____	20. 6 - 4 = ____	30. 8 - 6 = ____

Today, I finished _____ problems.

I solved _____ problems correctly.

Name _____   Date _____

My Mixed Practice

1. $4 + 2 = $ ___	11. $2 + $ ___ $ = 6$	21. $8 - 5 = $ ___
2. $2 + $ ___ $ = 6$	12. $6 - 2 = $ ___	22. $3 + $ ___ $ = 8$
3. $6 = 3 + $ ___	13. $6 - 4 = $ ___	23. $8 = $ ___ $ + 5$
4. $2 + 5 = $ ___	14. $5 + $ ___ $ = 7$	24. ___ $ + 2 = 9$
5. $7 = 5 + $ ___	15. $7 - 5 = $ ___	25. $9 = $ ___ $ + 7$
6. $4 + 3 = $ ___	16. $7 - 4 = $ ___	26. $9 - 2 = $ ___
7. $7 = $ ___ $ + 4$	17. $7 - 3 = $ ___	27. $9 - 7 = $ ___
8. $8 = $ ___ $ + 4$	18. $8 = 6 + $ ___	28. $9 - 6 = $ ___
9. $4 + 5 = $ ___	19. $8 - 2 = $ ___	29. $9 = $ ___ $ + 4$
10. $9 = $ ___ $ + 4$	20. $8 - 6 = $ ___	30. $9 - 6 = $ ___

Today, I finished _____ problems.

I solved _____ problems correctly.

A

Name _____

Date _____

Number Correct: ⬟

*Write the missing number.

1.	$5 + \square = 10$		16.	$9 + \square = 10$	
2.	$9 + \square = 10$		17.	$19 + \square = 20$	
3.	$10 + \square = 10$		18.	$5 + \square = 10$	
4.	$0 + \square = 10$		19.	$15 + \square = 20$	
5.	$8 + \square = 10$		20.	$1 + \square = 10$	
6.	$7 + \square = 10$		21.	$11 + \square = 20$	
7.	$6 + \square = 10$		22.	$3 + \square = 10$	
8.	$4 + \square = 10$		23.	$13 + \square = 20$	
9.	$3 + \square = 10$		24.	$4 + \square = 10$	
10.	$\square + 7 = 10$		25.	$14 + \square = 20$	
11.	$2 + \square = 10$		26.	$16 + \square = 20$	
12.	$\square + 8 = 10$		27.	$2 + \square = 10$	
13.	$1 + \square = 10$		28.	$12 + \square = 20$	
14.	$\square + 2 = 10$		29.	$18 + \square = 20$	
15.	$\square + 3 = 10$		30.	$11 + \square = 20$	

EUREKA
MATH™

Lesson 25: Add a pair of two-digit numbers when the ones digits have a sum less than or equal to 10.

47

B

Number Correct:

Name _____   Date _____

*Write the missing number.

1.	$10 + \square = 10$		16.	$5 + \square = 10$	
2.	$0 + \square = 10$		17.	$15 + \square = 20$	
3.	$9 + \square = 10$		18.	$9 + \square = 10$	
4.	$5 + \square = 10$		19.	$19 + \square = 20$	
5.	$6 + \square = 10$		20.	$8 + \square = 10$	
6.	$7 + \square = 10$		21.	$18 + \square = 20$	
7.	$8 + \square = 10$		22.	$2 + \square = 10$	
8.	$2 + \square = 10$		23.	$12 + \square = 20$	
9.	$3 + \square = 10$		24.	$3 + \square = 10$	
10.	$\square + 7 = 10$		25.	$13 + \square = 20$	
11.	$2 + \square = 10$		26.	$17 + \square = 20$	
12.	$\square + 8 = 10$		27.	$4 + \square = 10$	
13.	$1 + \square = 10$		28.	$16 + \square = 20$	
14.	$\square + 9 = 10$		29.	$18 + \square = 20$	
15.	$\square + 2 = 10$		30.	$12 + \square = 40$	

Names _____   Date _____

 Race to the Top!

2	3	4	5	6	7	8	9	10	11	12

_____

race to the top

Lesson 27:   Add a pair of two-digit numbers when the ones digits have a sum greater than 10.

51

©2015 Great Minds®. eureka-math.org

Names _____   Date _____

 Race to the Top!

2	**3**	**4**	**5**	**6**	**7**	**8**	**9**	**10**	**11**	**12**

_____

race to the top

Grade 1
Module 5

A

Name _____

Number Correct:

Date _____

*Write the unknown number. Pay attention to the symbols.

1.	4 + 1 = _____	16.	4 + 3 = _____
2.	4 + 2 = _____	17.	_____ + 4 = 7
3.	4 + 3 = _____	18.	7 = _____ + 4
4.	6 + 1 = _____	19.	5 + 4 = _____
5.	6 + 2 = _____	20.	_____ + 5 = 9
6.	6 + 3 = _____	21.	9 = _____ + 4
7.	1 + 5 = _____	22.	2 + 7 = _____
8.	2 + 5 = _____	23.	_____ + 2 = 9
9.	3 + 5 = _____	24.	9 = _____ + 7
10.	5 + _____ = 8	25.	3 + 6 = _____
11.	8 = 3 + _____	26.	_____ + 3 = 9
12.	7 + 2 = _____	27.	9 = _____ + 6
13.	7 + 3 = _____	28.	4 + 4 = _____ + 2
14.	7 + _____ = 10	29.	5 + 4 = _____ + 3
15.	_____ + 7 = 10	30.	_____ + 7 = 3 + 6

B

Name _____

Number Correct:

Date _____

*Write the unknown number. Pay attention to the symbols.

1.	5 + 1 = _____	16.	2 + 4 = _____
2.	5 + 2 = _____	17.	_____ + 4 = 6
3.	5 + 3 = _____	18.	6 = _____ + 4
4.	4 + 1 = _____	19.	3 + 4 = _____
5.	4 + 2 = _____	20.	_____ + 3 = 7
6.	4 + 3 = _____	21.	7 = _____ + 4
7.	1 + 3 = _____	22.	4 + 5 = _____
8.	2 + 3 = _____	23.	_____ + 4 = 9
9.	3 + 3 = _____	24.	9 = _____ + 5
10.	3 + _____ = 6	25.	2 + 6 = _____
11.	_____ + 3 = 6	26.	_____ + 6 = 9
12.	5 + 2 = _____	27.	9 = _____ + 2
13.	5 + 3 = _____	28.	3 + 3 = _____ + 4
14.	5 + _____ = 8	29.	3 + 4 = _____ + 5
15.	_____ + 3 = 8	30.	_____ + 6 = 2 + 7

Lesson 1:   Classify shapes based on defining attributes using examples, variants, and non-examples.

59

A

Name _____

Number Correct:

Date _____

*Write the unknown number. Pay attention to the equal sign.

1.	5 + 2 = ____	16.	____ = 5 + 4
2.	6 + 2 = ____	17.	____ = 4 + 5
3.	7 + 2 = ____	18.	6 + 3 = ____
4.	4 + 3 = ____	19.	3 + 6 = ____
5.	5 + 3 = ____	20.	____ = 2 + 6
6.	6 + 3 = ____	21.	2 + 7 = ____
7.	____ = 6 + 2	22.	____ = 3 + 4
8.	____ = 2 + 6	23.	3 + 6 = ____
9.	____ = 7 + 2	24.	____ = 4 + 5
10.	____ = 2 + 7	25.	3 + 4 = ____
11.	____ = 4 + 3	26.	13 + 4 = ____
12.	____ = 3 + 4	27.	3 + 14 = ____
13.	____ = 5 + 3	28.	3 + 6 = ____
14.	____ = 3 + 5	29.	13 + ____ = 19
15.	____ = 3 + 4	30.	19 = ____ + 16

EUREKA MATH™

Lesson 1:   Classify shapes based on defining attributes using examples, variants, and non-examples.

61

B

Name _____   Date _____

Number Correct: ⭐

*Write the unknown number. Pay attention to the equal sign.

1.	4 + 3 = ____	16.	____ = 6 + 3
2.	5 + 3 = ____	17.	____ = 3 + 6
3.	6 + 3 = ____	18.	5 + 4 = ____
4.	6 + 2 = ____	19.	4 + 5 = ____
5.	7 + 2 = ____	20.	____ = 2 + 7
6.	5 + 4 = ____	21.	2 + 6 = ____
7.	____ = 4 + 3	22.	____ = 3 + 4
8.	____ = 3 + 4	23.	4 + 5 = ____
9.	____ = 5 + 3	24.	____ = 3 + 6
10.	____ = 3 + 5	25.	2 + 7 = ____
11.	____ = 6 + 2	26.	12 + 7 = ____
12.	____ = 2 + 6	27.	2 + 17 = ____
13.	____ = 7 + 2	28.	4 + 5 = ____
14.	____ = 2 + 7	29.	14 + ____ = 19
15.	____ = 7 + 2	30.	19 = ____ + 15

Lesson 1:   Classify shapes based on defining attributes using examples, variants, and non-examples.

A

Name _____

Number Correct:

Date _____

*Write the unknown number. Pay attention to the symbols.

1.	6 - 1 = ____	16.	8 - 2 = ____
2.	6 - 2 = ____	17.	8 - 6 = ____
3.	6 - 3 = ____	18.	7 - 3 = ____
4.	10 - 1 = ____	19.	7 - 4 = ____
5.	10 - 2 = ____	20.	8 - 4 = ____
6.	10 - 3 = ____	21.	9 - 4 = ____
7.	7 - 2 = ____	22.	9 - 5 = ____
8.	8 - 2 = ____	23.	9 - 6 = ____
9.	9 - 2 = ____	24.	9 - ____ = 6
10.	7 - 3 = ____	25.	9 - ____ = 2
11.	8 - 3 = ____	26.	2 = 8 - ____
12.	10 - 3 = ____	27.	2 = 9 - ____
13.	10 - 4 = ____	28.	10 - 7 = 9 - ____
14.	9 - 4 = ____	29.	9 - 5 = ____ - 3
15.	8 - 4 = ____	30.	____ - 6 = 9 - 7

Lesson 1:   Classify shapes based on defining attributes using examples, variants, and non-examples.

65

B

Name _____

Number Correct:

Date _____

*Write the unknown number. Pay attention to the symbols.

1.	5 - 1 = _____	16.	6 - 2 = _____
2.	5 - 2 = _____	17.	6 - 4 = _____
3.	5 - 3 = _____	18.	8 - 3 = _____
4.	10 - 1 = _____	19.	8 - 5 = _____
5.	10 - 2 = _____	20.	8 - 6 = _____
6.	10 - 3 = _____	21.	9 - 3 = _____
7.	6 - 2 = _____	22.	9 - 6 = _____
8.	7 - 2 = _____	23.	9 - 7 = _____
9.	8 - 2 = _____	24.	9 - _____ = 5
10.	6 - 3 = _____	25.	9 - _____ = 4
11.	7 - 3 = _____	26.	4 = 8 - _____
12.	8 - 3 = _____	27.	4 = 9 - _____
13.	5 - 4 = _____	28.	10 - 8 = 9 - _____
14.	6 - 4 = _____	29.	8 - 6 = _____ - 7
15.	7 - 4 = _____	30.	_____ - 4 = 9 - 6

Lesson 1:   Classify shapes based on defining attributes using examples, variants, and non-examples.

67

A

Name _____

Number Correct: ⬡

Date _____

*Write the unknown number. Pay attention to the symbols.

1.	$2 + 3 =$	16.	$3 + 3 =$
2.	$3 + \rule{1cm}{0.1mm} = 5$	17.	$6 - 3 =$
3.	$5 - 3 =$	18.	$6 = \rule{1cm}{0.1mm} + 3$
4.	$5 - 2 =$	19.	$2 + 5 =$
5.	$\rule{1cm}{0.1mm} + 2 = 5$	20.	$5 + \rule{1cm}{0.1mm} = 7$
6.	$1 + 5 =$	21.	$7 - 2 =$
7.	$1 + \rule{1cm}{0.1mm} = 6$	22.	$7 - 5 =$
8.	$6 - 1 =$	23.	$7 = \rule{1cm}{0.1mm} + 5$
9.	$6 - 5 =$	24.	$3 + 4 =$
10.	$\rule{1cm}{0.1mm} + 5 = 6$	25.	$4 + \rule{1cm}{0.1mm} = 7$
11.	$4 + 2 =$	26.	$7 - 4 =$
12.	$2 + \rule{1cm}{0.1mm} = 6$	27.	$7 = \rule{1cm}{0.1mm} + 3$
13.	$6 - 2 =$	28.	$3 = 7 - \rule{1cm}{0.1mm}$
14.	$6 - 4 =$	29.	$7 - 5 = \rule{1cm}{0.1mm} - 4$
15.	$\rule{1cm}{0.1mm} + 4 = 6$	30.	$\rule{1cm}{0.1mm} - 3 = 7 - 4$

B

Number Correct:

Name _____   Date _____

*Write the unknown number. Pay attention to the symbols.

1.	1 + 4 =	16.	3 + 3 =
2.	4 + _____ = 5	17.	6 - 3 =
3.	5 - 4 =	18.	6 = _____ + 3
4.	5 - 1 =	19.	2 + 4 =
5.	+ 1 = 5	20.	4 + _____ = 6
6.	5 + 2 =	21.	6 - 2 =
7.	5 + _____ = 7	22.	6 - 4 =
8.	7 - 2 =	23.	6 = _____ + 4
9.	7 - 5 = _____	24.	3 + 4 =
10.	+ 2 = 7	25.	4 + _____ = 7
11.	1 + 5 =	26.	7 - 4 =
12.	1 + _____ = 6	27.	7 = _____ + 4
13.	6 - 1 =	28.	4 = 7 -
14.	6 - 5 =	29.	6 - 4 = _____ - 5
15.	+ 5 = 6	30.	- 2 = 7 - 3

A

Name _____

Number Correct:

Date _____

*Write the unknown number. Pay attention to the symbols.

1.	$5 + 5 =$	16.	$2 + 6 =$
2.	$5 + \underline{\quad} = 10$	17.	$8 = 6 +$
3.	$10 - 5 =$	18.	$8 - 2 =$
4.	$9 + 1 =$	19.	$2 + 7 =$
5.	$1 + \underline{\quad} = 10$	20.	$9 = 7 +$
6.	$10 - 1 =$	21.	$9 - 7 =$
7.	$10 - 9 =$	22.	$8 = \underline{\quad} + 2$
8.	$+ 9 = 10$	23.	$8 - 6 =$
9.	$1 + 8 =$	24.	$3 + 6 =$
10.	$8 + \underline{\quad} = 9$	25.	$9 = 6 +$
11.	$9 - 1 =$	26.	$9 - 6 =$
12.	$9 - 8 =$	27.	$9 = \underline{\quad} + 3$
13.	$+ 1 = 9$	28.	$3 = 9 -$
14.	$4 + 4 =$	29.	$9 - 5 = \underline{\quad} - 6$
15.	$8 - 4 =$	30.	$- 7 = 8 - 6$

B

Name _____

Number Correct:

Date _____

*Write the unknown number. Pay attention to the symbols.

1.	9 + 1 =	16.	3 + 5 =
2.	1 + _____ = 10	17.	8 = 5 +
3.	10 - 1 =	18.	8 - 3 =
4.	10 - 9 =	19.	2 + 6 =
5.	+ 9 = 10	20.	8 = 6 +
6.	1 + 7 =	21.	8 - 6 =
7.	7 + _____ = 8	22.	2 + 7 =
8.	8 - 1 =	23.	9 = _____ + 2
9.	8 - 7 =	24.	9 - 7 =
10.	+ 1 = 8	25.	4 + 5 =
11.	2 + 8 =	26.	9 = 5 +
12.	2 + _____ = 10	27.	9 - 5 =
13.	10 - 2 =	28.	5 = 9 -
14.	10 - 8 =	29.	9 - 6 = _____ - 5
15.	+ 8 = 10	30.	- 6 = 9 - 7

EUREKA
MATH™

Name _____   Date _____

My Addition Practice

1. 6 + 0 = ___	11. 7 + 1 = ___	21. 5 + 3 = ___
2. 0 + 6 = ___	12. ___ = 1 + 7	22. ___ = 5 + 4
3. 5 + 1 = ___	13. 3 + 3 = ___	23. 6 + 4 = ___
4. 1 + 5 = ___	14. 3 + 4 = ___	24. 4 + 6 = ___
5. 6 + 1 = ___	15. ___ = 3 + 5	25. ___ = 4 + 4
6. 1 + 6 = ___	16. 6 + 3 = ___	26. 3 + 4 = ___
7. 6 + 2 = ___	17. 7 + 3 = ___	27. 5 + 5 = ___
8. 5 + 2 = ___	18. ___ = 7 + 2	28. ___ = 4 + 5
9. 2 + 5 = ___	19. 2 + 7 = ___	29. 3 + 7 = ___
10. 2 + 4 = ___	20. 2 + 8 = ___	30. ___ = 3 + 6

Today, I finished _____ problems.

EUREKA MATH

Lesson 3:   Find and name three-dimensional shapes including cone and rectangular prism, based on defining attributes of faces and points.

77

Name _____   Date _____

My Missing Addend Practice

1. $6 + \underline{\hphantom{0}} = 6$

2. $0 + \underline{\hphantom{0}} = 6$

3. $5 + \underline{\hphantom{0}} = 6$

4. $4 + \underline{\hphantom{0}} = 6$

5. $0 + \underline{\hphantom{0}} = 7$

6. $6 + \underline{\hphantom{0}} = 7$

7. $1 + \underline{\hphantom{0}} = 7$

8. $7 + \underline{\hphantom{0}} = 8$

9. $1 + \underline{\hphantom{0}} = 8$

10. $6 + \underline{\hphantom{0}} = 8$

11. $3 + \underline{\hphantom{0}} = 6$

12. $4 + \underline{\hphantom{0}} = 8$

13. $10 = 5 + \underline{\hphantom{0}}$

14. $5 + \underline{\hphantom{0}} = 9$

15. $5 + \underline{\hphantom{0}} = 7$

16. $8 = 5 + \underline{\hphantom{0}}$

17. $5 + \underline{\hphantom{0}} = 9$

18. $8 + \underline{\hphantom{0}} = 10$

19. $7 + \underline{\hphantom{0}} = 10$

20. $10 = 6 + \underline{\hphantom{0}}$

21. $4 + \underline{\hphantom{0}} = 7$

22. $7 = 3 + \underline{\hphantom{0}}$

23. $2 + \underline{\hphantom{0}} = 7$

24. $2 + \underline{\hphantom{0}} = 8$

25. $9 = 2 + \underline{\hphantom{0}}$

26. $2 + \underline{\hphantom{0}} = 10$

27. $10 = 3 + \underline{\hphantom{0}}$

28. $3 + \underline{\hphantom{0}} = 9$

29. $4 + \underline{\hphantom{0}} = 9$

30. $10 = 4 + \underline{\hphantom{0}}$

Today, I finished _____ problems.

I solved _____ problems correctly.

EUREKA MATH   Lesson 3: Find and name three-dimensional shapes including cone and rectangular prism, based on defining attributes of faces and points.   79

©2015 Great Minds®. eureka-math.org

Name _____   Date _____

My Related Addition and Subtraction Practice

1. $5 + \_\_\_ = 6$

2. $1 + \_\_\_ = 6$

3. $6 - 1 = \_\_\_$

4. $9 + \_\_\_ = 10$

5. $1 + \_\_\_ = 10$

6. $10 - 9 = \_\_\_$

7. $5 + \_\_\_ = 10$

8. $10 - 5 = \_\_\_$

9. $8 + \_\_\_ = 10$

10. $10 - 8 = \_\_\_$

11. $7 + \_\_\_ = 10$

12. $10 - 7 = \_\_\_$

13. $5 + \_\_\_ = 7$

14. $7 - 5 = \_\_\_$

15. $5 + \_\_\_ = 8$

16. $8 - 5 = \_\_\_$

17. $4 + \_\_\_ = 6$

18. $6 - 4 = \_\_\_$

19. $3 + \_\_\_ = 6$

20. $6 - 3 = \_\_\_$

21. $4 + \_\_\_ = 8$

22. $8 - 4 = \_\_\_$

23. $4 + \_\_\_ = 7$

24. $7 - 4 = \_\_\_$

25. $5 + \_\_\_ = 9$

26. $9 - 5 = \_\_\_$

27. $6 + \_\_\_ = 9$

28. $9 - 6 = \_\_\_$

29. $4 + \_\_\_ = 7$

30. $7 - 4 = \_\_\_$

Today, I finished _____ problems.

I solved _____ problems correctly.

EUREKA MATH™

Lesson 3:   Find and name three-dimensional shapes including cone and rectangular prism, based on defining attributes of faces and points.

Name _____   Date _____

My Subtraction Practice

1. $6 - 0 =$ ___   11. $6 - 3 =$ ___   21. $8 - 4 =$ ___

2. $6 - 1 =$ ___   12. $7 - 3 =$ ___   22. $8 - 3 =$ ___

3. $7 - 1 =$ ___   13. $9 - 3 =$ ___   23. $8 - 5 =$ ___

4. $8 - 1 =$ ___   14. $10 - 8 =$ ___   24. $9 - 5 =$ ___

5. $6 - 2 =$ ___   15. $10 - 6 =$ ___   25. $9 - 4 =$ ___

6. $7 - 2 =$ ___   16. $10 - 4 =$ ___   26. $7 - 3 =$ ___

7. $9 - 2 =$ ___   17. $10 - 5 =$ ___   27. $10 - 7 =$ ___

8. $10 - 10 =$ ___   18. $7 - 6 =$ ___   28. $9 - 7 =$ ___

9. $10 - 9 =$ ___   19. $7 - 5 =$ ___   29. $9 - 6 =$ ___

10. $10 - 7 =$ ___   20. $6 - 4 =$ ___   30. $8 - 6 =$ ___

Today, I finished _____ problems.

I solved _____ problems correctly.

EUREKA MATH™

Lesson 3:   Find and name three-dimensional shapes including cone and rectangular prism, based on defining attributes of faces and points.

83

©2015 Great Minds®. eureka-math.org

Name _____   Date _____

My Mixed Practice

1. $4 + 2 =$ ___	11. $2 +$ ___ $= 6$	21. $8 - 5 =$ ___
2. $2 +$ ___ $= 6$	12. $6 - 2 =$ ___	22. $3 +$ ___ $= 8$
3. $6 = 3 +$ ___	13. $6 - 4 =$ ___	23. $8 =$ ___ $+ 5$
4. $2 + 5 =$ ___	14. $5 +$ ___ $= 7$	24. ___ $+ 2 = 9$
5. $7 = 5 +$ ___	15. $7 - 5 =$ ___	25. $9 =$ ___ $+ 7$
6. $4 + 3 =$ ___	16. $7 - 4 =$ ___	26. $9 - 2 =$ ___
7. $7 =$ ___ $+ 4$	17. $7 - 3 =$ ___	27. $9 - 7 =$ ___
8. $8 =$ ___ $+ 4$	18. $8 = 6 +$ ___	28. $9 - 6 =$ ___
9. $4 + 5 =$ ___	19. $8 - 2 =$ ___	29. $9 =$ ___ $+ 4$
10. $9 =$ ___ $+ 4$	20. $8 - 6 =$ ___	30. $9 - 6 =$ ___

Today, I finished _____ problems.

I solved _____ problems correctly.

Grade 1
Module 6

Name _____   Date _____

My Addition Practice

1. $6 + 0 =$ ___	11. $7 + 1 =$ ___	21. $5 + 3 =$ ___
2. $0 + 6 =$ ___	12. ___ $= 1 + 7$	22. ___ $= 5 + 4$
3. $5 + 1 =$ ___	13. $3 + 3 =$ ___	23. $6 + 4 =$ ___
4. $1 + 5 =$ ___	14. $3 + 4 =$ ___	24. $4 + 6 =$ ___
5. $6 + 1 =$ ___	15. ___ $= 3 + 5$	25. ___ $= 4 + 4$
6. $1 + 6 =$ ___	16. $6 + 3 =$ ___	26. $3 + 4 =$ ___
7. $6 + 2 =$ ___	17. $7 + 3 =$ ___	27. $5 + 5 =$ ___
8. $5 + 2 =$ ___	18. ___ $= 7 + 2$	28. ___ $= 4 + 5$
9. $2 + 5 =$ ___	19. $2 + 7 =$ ___	29. $3 + 7 =$ ___
10. $2 + 4 =$ ___	20. $2 + 8 =$ ___	30. ___ $= 3 + 6$

Today I finished _____ problems.

I solved _____ problems correctly.

Name _____   Date _____

My Missing Addend Practice

1. $6 + \underline{\quad} = 6$	11. $3 + \underline{\quad} = 6$	21. $4 + \underline{\quad} = 7$
2. $0 + \underline{\quad} = 6$	12. $4 + \underline{\quad} = 8$	22. $7 = 3 + \underline{\quad}$
3. $5 + \underline{\quad} = 6$	13. $10 = 5 + \underline{\quad}$	23. $2 + \underline{\quad} = 7$
4. $4 + \underline{\quad} = 6$	14. $5 + \underline{\quad} = 9$	24. $2 + \underline{\quad} = 8$
5. $0 + \underline{\quad} = 7$	15. $5 + \underline{\quad} = 7$	25. $9 = 2 + \underline{\quad}$
6. $6 + \underline{\quad} = 7$	16. $8 = 5 + \underline{\quad}$	26. $2 + \underline{\quad} = 10$
7. $1 + \underline{\quad} = 7$	17. $5 + \underline{\quad} = 9$	27. $10 = 3 + \underline{\quad}$
8. $7 + \underline{\quad} = 8$	18. $8 + \underline{\quad} = 10$	28. $3 + \underline{\quad} = 9$
9. $1 + \underline{\quad} = 8$	19. $7 + \underline{\quad} = 10$	29. $4 + \underline{\quad} = 9$
10. $6 + \underline{\quad} = 8$	20. $10 = 6 + \underline{\quad}$	30. $10 = 4 + \underline{\quad}$

Today I finished _____ problems.

I solved _____ problems correctly.

Name _____   Date _____

My Related Addition and Subtraction Practice

1. $5 + \underline{\ \ } = 6$	11. $7 + \underline{\ \ } = 10$	21. $4 + \underline{\ \ } = 8$
2. $1 + \underline{\ \ } = 6$	12. $10 - 7 = \underline{\ \ }$	22. $8 - 4 = \underline{\ \ }$
3. $6 - 1 = \underline{\ \ }$	13. $5 + \underline{\ \ } = 7$	23. $4 + \underline{\ \ } = 7$
4. $9 + \underline{\ \ } = 10$	14. $7 - 5 = \underline{\ \ }$	24. $7 - 4 = \underline{\ \ }$
5. $1 + \underline{\ \ } = 10$	15. $5 + \underline{\ \ } = 8$	25. $5 + \underline{\ \ } = 9$
6. $10 - 9 = \underline{\ \ }$	16. $8 - 5 = \underline{\ \ }$	26. $9 - 5 = \underline{\ \ }$
7. $5 + \underline{\ \ } = 10$	17. $4 + \underline{\ \ } = 6$	27. $6 + \underline{\ \ } = 9$
8. $10 - 5 = \underline{\ \ }$	18. $6 - 4 = \underline{\ \ }$	28. $9 - 6 = \underline{\ \ }$
9. $8 + \underline{\ \ } = 10$	19. $3 + \underline{\ \ } = 6$	29. $4 + \underline{\ \ } = 7$
10. $10 - 8 = \underline{\ \ }$	20. $6 - 3 = \underline{\ \ }$	30. $7 - 4 = \underline{\ \ }$

Today I finished _____ problems.

I solved _____ problems correctly.

Name _____   Date _____

My Subtraction Practice

1. 6 – 0 = ___	11. 6 – 3 = ___	21. 8 – 4 = ___
2. 6 – 1 = ___	12. 7 – 3 = ___	22. 8 – 3 = ___
3. 7 – 1 = ___	13. 9 – 3 = ___	23. 8 – 5 = ___
4. 8 – 1 = ___	14. 10 – 8 = ___	24. 9 – 5 = ___
5. 6 – 2 = ___	15. 10 – 6 = ___	25. 9 – 4 = ___
6. 7 – 2 = ___	16. 10 – 4 = ___	26. 7 – 3 = ___
7. 9 – 2 = ___	17. 10 – 5 = ___	27. 10 – 7 = ___
8. 10 – 10 = ___	18. 7 – 6 = ___	28. 9 – 7 = ___
9. 10 – 9 = ___	19. 7 – 5 = ___	29. 9 – 6 = ___
10. 10 – 7 = ___	20. 6 – 4 = ___	30. 8 – 6 = ___

Today I finished _____ problems.

I solved _____ problems correctly.

Name _____   Date _____

My Mixed Practice

1. $4 + 2 =$ ___	11. $2 +$ ___ $= 6$	21. $8 - 5 =$ ___
2. $2 +$ ___ $= 6$	12. $6 - 2 =$ ___	22. $3 +$ ___ $= 8$
3. $6 = 3 +$ ___	13. $6 - 4 =$ ___	23. $8 =$ ___ $+ 5$
4. $2 + 5 =$ ___	14. $5 +$ ___ $= 7$	24. ___ $+ 2 = 9$
5. $7 = 5 +$ ___	15. $7 - 5 =$ ___	25. $9 =$ ___ $+ 7$
6. $4 + 3 =$ ___	16. $7 - 4 =$ ___	26. $9 - 2 =$ ___
7. $7 =$ ___ $+ 4$	17. $7 - 3 =$ ___	27. $9 - 7 =$ ___
8. $8 =$ ___ $+ 4$	18. $8 = 6 +$ ___	28. $9 - 6 =$ ___
9. $4 + 5 =$ ___	19. $8 - 2 =$ ___	29. $9 =$ ___ $+ 4$
10. $9 =$ ___ $+ 4$	20. $8 - 6 =$ ___	30. $9 - 6 =$ ___

Today I finished _____ problems.

I solved _____ problems correctly.

A

Name _____   Date _____

Number Correct: ⟨star⟩

*Write the unknown number. Pay attention to the symbols.

1.	4 + 1 = _____	16.	4 + 3 = _____
2.	4 + 2 = _____	17.	_____ + 4 = 7
3.	4 + 3 = _____	18.	7 = _____ + 4
4.	6 + 1 = _____	19.	5 + 4 = _____
5.	6 + 2 = _____	20.	_____ + 5 = 9
6.	6 + 3 = _____	21.	9 = _____ + 4
7.	1 + 5 = _____	22.	2 + 7 = _____
8.	2 + 5 = _____	23.	_____ + 2 = 9
9.	3 + 5 = _____	24.	9 = _____ + 7
10.	5 + _____ = 8	25.	3 + 6 = _____
11.	8 = 3 + _____	26.	_____ + 3 = 9
12.	7 + 2 = _____	27.	9 = _____ + 6
13.	7 + 3 = _____	28.	4 + 4 = _____ + 2
14.	7 + _____ = 10	29.	5 + 4 = _____ + 3
15.	_____ + 7 = 10	30.	_____ + 7 = 3 + 6

EUREKA MATH™

Lesson 3:   Use the place value chart to record and name tens and ones within a
two-digit number up to 100.

99

B

Name _____   Date _____

Number Correct: ⎯⎯⎯

*Write the unknown number. Pay attention to the symbols.

1.	5 + 1 = _____	16.	2 + 4 = _____
2.	5 + 2 = _____	17.	_____ + 4 = 6
3.	5 + 3 = _____	18.	6 = _____ + 4
4.	4 + 1 = _____	19.	3 + 4 = _____
5.	4 + 2 = _____	20.	_____ + 3 = 7
6.	4 + 3 = _____	21.	7 = _____ + 4
7.	1 + 3 = _____	22.	4 + 5 = _____
8.	2 + 3 = _____	23.	_____ + 4 = 9
9.	3 + 3 = _____	24.	9 = _____ + 5
10.	3 + _____ = 6	25.	2 + 6 = _____
11.	_____ + 3 = 6	26.	_____ + 6 = 9
12.	5 + 2 = _____	27.	9 = _____ + 2
13.	5 + 3 = _____	28.	3 + 3 = _____ + 4
14.	5 + _____ = 8	29.	3 + 4 = _____ + 5
15.	_____ + 3 = 8	30.	_____ + 6 = 2 + 7

EUREKA MATH

Lesson 3:   Use the place value chart to record and name tens and ones within a two-digit number up to 100.

101

A

Name _____   Date _____

Number Correct: _____

*Write the unknown number. Pay attention to the equal sign.

1.	5 + 2 = ____	16.	____ = 5 + 4
2.	6 + 2 = ____	17.	____ = 4 + 5
3.	7 + 2 = ____	18.	6 + 3 = ____
4.	4 + 3 = ____	19.	3 + 6 = ____
5.	5 + 3 = ____	20.	____ = 2 + 6
6.	6 + 3 = ____	21.	2 + 7 = ____
7.	____ = 6 + 2	22.	____ = 3 + 4
8.	____ = 2 + 6	23.	3 + 6 = ____
9.	____ = 7 + 2	24.	____ = 4 + 5
10.	____ = 2 + 7	25.	3 + 4 = ____
11.	____ = 4 + 3	26.	13 + 4 = ____
12.	____ = 3 + 4	27.	3 + 14 = ____
13.	____ = 5 + 3	28.	3 + 6 = ____
14.	____ = 3 + 5	29.	13 + ____ = 19
15.	____ = 3 + 4	30.	19 = ____ + 16

Lesson 3:   Use the place value chart to record and name tens and ones within a
            two-digit number up to 100.

103

B

Name _____   Date _____

Number Correct: _____

*Write the unknown number. Pay attention to the equal sign.

1.	4 + 3 = ____	16.	____ = 6 + 3
2.	5 + 3 = ____	17.	____ = 3 + 6
3.	6 + 3 = ____	18.	5 + 4 = ____
4.	6 + 2 = ____	19.	4 + 5 = ____
5.	7 + 2 = ____	20.	____ = 2 + 7
6.	5 + 4 = ____	21.	2 + 6 = ____
7.	____ = 4 + 3	22.	____ = 3 + 4
8.	____ = 3 + 4	23.	4 + 5 = ____
9.	____ = 5 + 3	24.	____ = 3 + 6
10.	____ = 3 + 5	25.	2 + 7 = ____
11.	____ = 6 + 2	26.	12 + 7 = ____
12.	____ = 2 + 6	27.	2 + 17 = ____
13.	____ = 7 + 2	28.	4 + 5 = ____
14.	____ = 2 + 7	29.	14 + ____ = 19
15.	____ = 7 + 2	30.	19 = ____ + 15

EUREKA MATH   Lesson 3: Use the place value chart to record and name tens and ones within a two-digit number up to 100.   105

©2015 Great Minds®. eureka-math.org

A

Number Correct:

Name _____   Date _____

*Write the unknown number. Pay attention to the symbols.

1.	6 – 1 = _____	16.	8 – 2 = _____
2.	6 – 2 = _____	17.	8 – 6 = _____
3.	6 – 3 = _____	18.	7 – 3 = _____
4.	10 – 1 = _____	19.	7 – 4 = _____
5.	10 – 2 = _____	20.	8 – 4 = _____
6.	10 – 3 = _____	21.	9 – 4 = _____
7.	7 – 2 = _____	22.	9 – 5 = _____
8.	8 – 2 = _____	23.	9 – 6 = _____
9.	9 – 2 = _____	24.	9 – _____ = 6
10.	7 – 3 = _____	25.	9 – _____ = 2
11.	8 – 3 = _____	26.	2 = 8 – _____
12.	10 – 3 = _____	27.	2 = 9 – _____
13.	10 – 4 = _____	28.	10 – 7 = 9 – _____
14.	9 – 4 = _____	29.	9 – 5 = _____ – 3
15.	8 – 4 = _____	30.	_____ – 6 = 9 – 7

B

Number Correct: _____

Name _____   Date _____

*Write the unknown number. Pay attention to the symbols.

1.	5 – 1 = _____	16.	6 – 2 = _____
2.	5 – 2 = _____	17.	6 – 4 = _____
3.	5 – 3 = _____	18.	8 – 3 = _____
4.	10 – 1 = _____	19.	8 – 5 = _____
5.	10 – 2 = _____	20.	8 – 6 = _____
6.	10 – 3 = _____	21.	9 – 3 = _____
7.	6 – 2 = _____	22.	9 – 6 = _____
8.	7 – 2 = _____	23.	9 – 7 = _____
9.	8 – 2 = _____	24.	9 – _____ = 5
10.	6 – 3 = _____	25.	9 – _____ = 4
11.	7 – 3 = _____	26.	4 = 8 – _____
12.	8 – 3 = _____	27.	4 = 9 – _____
13.	5 – 4 = _____	28.	10 – 8 = 9 – _____
14.	6 – 4 = _____	29.	8 – 6 = _____ – 7
15.	7 – 4 = _____	30.	_____ – 4 = 9 – 6

Lesson 3:   Use the place value chart to record and name tens and ones within a
two-digit number up to 100.   **109**

A

Name _____   Date _____

Number Correct: ⬤

*Write the unknown number. Pay attention to the symbols.

1.	$2 + 3 = $ _____	16.	$3 + 3 = $ _____
2.	$3 + $ _____ $= 5$	17.	$6 - 3 = $ _____
3.	$5 - 3 = $ _____	18.	$6 = $ _____ $+ 3$
4.	$5 - 2 = $ _____	19.	$2 + 5 = $ _____
5.	_____ $+ 2 = 5$	20.	$5 + $ _____ $= 7$
6.	$1 + 5 = $ _____	21.	$7 - 2 = $ _____
7.	$1 + $ _____ $= 6$	22.	$7 - 5 = $ _____
8.	$6 - 1 = $ _____	23.	$7 = $ _____ $+ 5$
9.	$6 - 5 = $ _____	24.	$3 + 4 = $ _____
10.	_____ $+ 5 = 6$	25.	$4 + $ _____ $= 7$
11.	$4 + 2 = $ _____	26.	$7 - 4 = $ _____
12.	$2 + $ _____ $= 6$	27.	$7 = $ _____ $+ 3$
13.	$6 - 2 = $ _____	28.	$3 = 7 - $ _____
14.	$6 - 4 = $ _____	29.	$7 - 5 = $ _____ $- 4$
15.	_____ $+ 4 = 6$	30.	_____ $- 3 = 7 - 4$

B

Number Correct:

Name _____   Date _____

*Write the unknown number. Pay attention to the symbols.

1.	$1 + 4 = $ ____	16.	$3 + 3 = $ ____
2.	$4 + $ ____ $ = 5$	17.	$6 - 3 = $ ____
3.	$5 - 4 = $ ____	18.	$6 = $ ____ $ + 3$
4.	$5 - 1 = $ ____	19.	$2 + 4 = $ ____
5.	____ $ + 1 = 5$	20.	$4 + $ ____ $ = 6$
6.	$7 + 2 = $ ____	21.	$6 - 2 = $ ____
7.	$5 + $ ____ $ = 7$	22.	$6 - 4 = $ ____
8.	$7 - 2 = $ ____	23.	$6 = $ ____ $ + 4$
9.	$7 - 5 = $ ____	24.	$3 + 4 = $ ____
10.	____ $ + 2 = 7$	25.	$4 + $ ____ $ = 7$
11.	$1 + 5 = $ ____	26.	$7 - 4 = $ ____
12.	$1 + $ ____ $ = 6$	27.	$7 = $ ____ $ + 4$
13.	$6 - 1 = $ ____	28.	$4 = 7 - $ ____
14.	$6 - 5 = $ ____	29.	$6 - 4 = $ ____ $ - 5$
15.	____ $ + 5 = 6$	30.	____ $ - 4 = 7 - 3$

A

Number Correct: _____

Name _____   Date _____

*Write the unknown number. Pay attention to the symbols.

1.	5 + 5 = _____	16.	2 + 6 = _____
2.	5 + _____ = 10	17.	8 = 6 + _____
3.	10 – 5 = _____	18.	8 – 2 = _____
4.	9 + 1 = _____	19.	2 + 7 = _____
5.	1 + _____ = 10	20.	9 = 7 + _____
6.	10 – 1 = _____	21.	9 – 7 = _____
7.	10 – 9 = _____	22.	8 = _____ + 2
8.	_____ + 9 = 10	23.	8 – 6 = _____
9.	1 + 8 = _____	24.	3 + 6 = _____
10.	8 + _____ = 9	25.	9 = 6 + _____
11.	9 – 1 = _____	26.	9 – 6 = _____
12.	9 – 8 = _____	27.	9 = _____ + 3
13.	_____ + 1 = 9	28.	3 = 9 – _____
14.	4 + 4 = _____	29.	9 – 5 = _____ – 6
15.	8 – 4 = _____	30.	_____ – 7 = 8 – 6

EUREKA MATH

Lesson 3:   Use the place value chart to record and name tens and ones within a two-digit number up to 100.

115

B

Number Correct: _____

Name _____   Date _____

*Write the unknown number. Pay attention to the symbols.

1.	9 + 1 = _____	16.	3 + 5 = _____
2.	1 + _____ = 10	17.	8 = 5 + _____
3.	10 − 1 = _____	18.	8 − 3 = _____
4.	10 − 9 = _____	19.	2 + 6 = _____
5.	_____ + 9 = 10	20.	8 = 6 + _____
6.	1 + 7 = _____	21.	8 − 6 = _____
7.	7 + _____ = 8	22.	2 + 7 = _____
8.	8 − 1 = _____	23.	9 = _____ + 2
9.	8 − 7 = _____	24.	9 − 7 = _____
10.	_____ + 1 = 8	25.	4 + 5 = _____
11.	2 + 8 = _____	26.	9 = 5 + _____
12.	2 + _____ = 10	27.	9 − 5 = _____
13.	10 − 2 = _____	28.	5 = 9 − _____
14.	10 − 8 = _____	29.	9 − 6 = _____ − 5
15.	_____ + 8 = 10	30.	_____ − 6 = 9 − 7

EUREKA
MATH™

Lesson 3:   Use the place value chart to record and name tens and ones within a
two-digit number up to 100.

117

A

Name _____   Date _____

Number Correct:

*Write the missing number. Pay attention to the addition or subtraction sign.

1.	5 + 1 = ☐		16.	29 + 10 = ☐	
2.	15 + 1 = ☐		17.	9 + 1 = ☐	
3.	25 + 1 = ☐		18.	19 + 1 = ☐	
4.	5 + 10 = ☐		19.	29 + 1 = ☐	
5.	15 + 10 = ☐		20.	39 + 1 = ☐	
6.	25 + 10 = ☐		21.	40 – 1 = ☐	
7.	8 – 1 = ☐		22.	30 – 1 = ☐	
8.	18 – 1 = ☐		23.	20 – 1 = ☐	
9.	28 – 1 = ☐		24.	20 + ☐ = 21	
10.	38 – 1 = ☐		25.	20 + ☐ = 30	
11.	38 – 10 = ☐		26.	27 + ☐ = 37	
12.	28 – 10 = ☐		27.	27 + ☐ = 28	
13.	18 – 10 = ☐		28.	☐ + 10 = 34	
14.	9 + 10 = ☐		29.	☐ – 10 = 14	
15.	19 + 10 = ☐		30.	☐ – 10 = 24	

EUREKA MATH™

Lesson 9:   Represent up to 120 objects with a written numeral.

119

B

Number Correct: _____

Name _____ Date _____

*Write the missing number. Pay attention to the addition or subtraction sign.

1.	4 + 1 = ☐		16.	28 + 10 = ☐	
2.	14 + 1 = ☐		17.	9 + 1 = ☐	
3.	24 + 1 = ☐		18.	19 + 1 = ☐	
4.	6 + 10 = ☐		19.	29 + 1 = ☐	
5.	16 + 10 = ☐		20.	39 + 1 = ☐	
6.	26 + 10 = ☐		21.	40 – 1 = ☐	
7.	7 – 1 = ☐		22.	30 – 1 = ☐	
8.	17 – 1 = ☐		23.	20 – 1 = ☐	
9.	27 – 1 = ☐		24.	10 + ☐ = 11	
10.	37 – 1 = ☐		25.	10 + ☐ = 20	
11.	37 – 10 = ☐		26.	22 + ☐ = 32	
12.	27 – 10 = ☐		27.	22 + ☐ = 23	
13.	17 – 10 = ☐		28.	☐ + 10 = 39	
14.	8 + 10 = ☐		29.	☐ – 10 = 19	
15.	18 + 10 = ☐		30.	☐ – 10 = 29	

Lesson 9: Represent up to 120 objects with a written numeral.

121

EUREKA
MATH™

Names _____   Date _____

 Race to the Top!

2	3	4	5	6	7	8	9	10	11	12

_____

race to the top

Name _____

Partner _____

Example

Step 1: Rewrite 4 – 1 as 1 + ____ = 4.

Step 2: Exchange papers and solve.

List A

1. 10 – 9 _____

2. 10 – 8 _____

3. 9 – 8 _____

4. 9 – 6 _____

5. 8 – 6 _____

6. 7 – 4 _____

7. 7 – 5 _____

8. 8 – 5 _____

9. 9 – 5 _____

10. 9 – 6 _____

Name _____

Partner _____

Example

Step 1: Rewrite 4 – 1 as 1 + ____ = 4.

Step 2: Exchange papers and solve.

List B

1. 10 – 8 _____

2. 10 – 7 _____

3. 8 – 7 _____

4. 8 – 6 _____

5. 9 – 6 _____

6. 7 – 6 _____

7. 7 – 5 _____

8. 7 – 4 _____

9. 8 – 5 _____

10. 6 – 4 _____

_____

pattern sheet list A or B

 EUREKA MATH™   **Lesson 18:**   Add a pair of two-digit numbers with varied sums in the ones, and compare the results of different recording methods.   125

©2015 Great Minds®. eureka-math.org

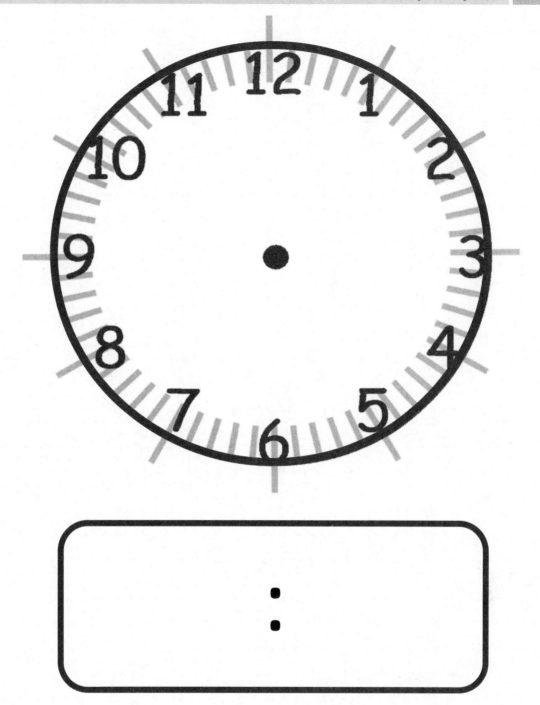

It is ___ o'clock.   It is half past ___.

_____
time recording sheet

Lesson 26: Solve compare with bigger or smaller unknown problem types.

127

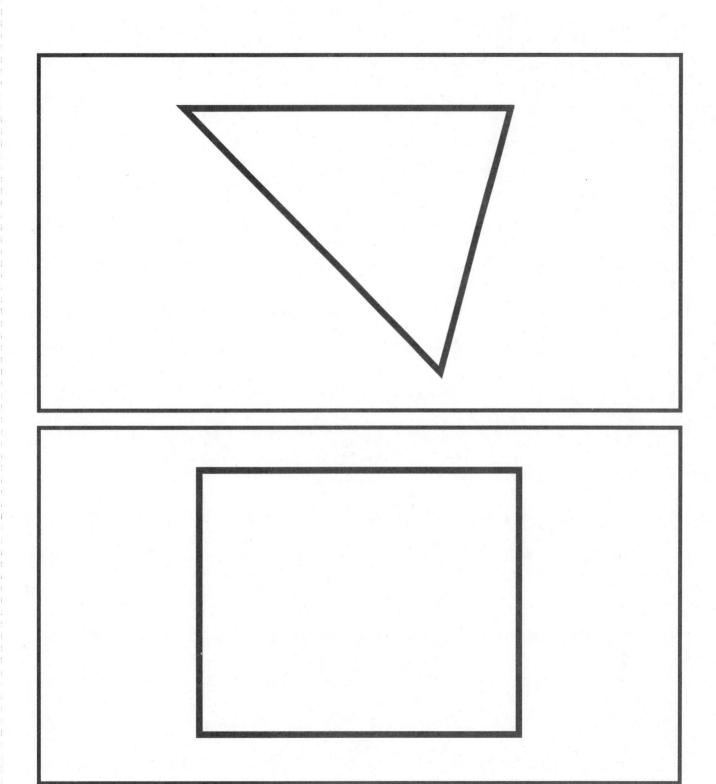

two-dimensional shape flashcards

Lesson 27:   Share and critique peer strategies for solving problems of varied types.   **129**

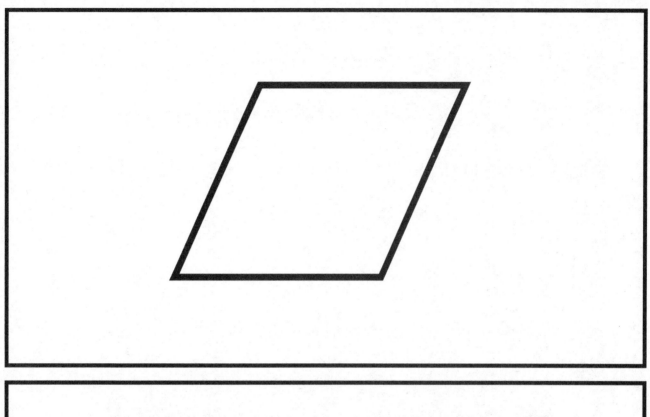

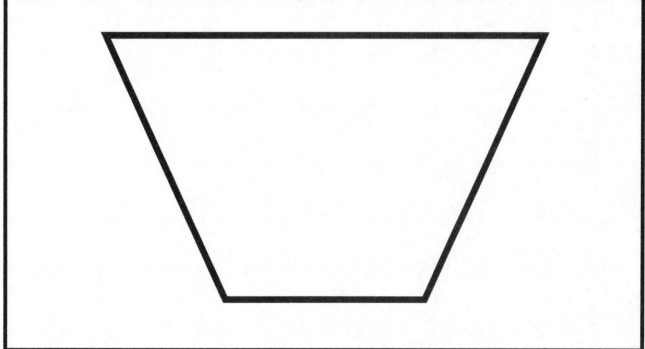

two-dimensional shape flashcards

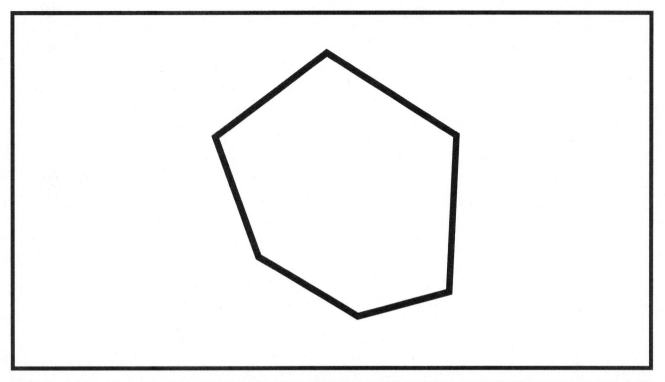

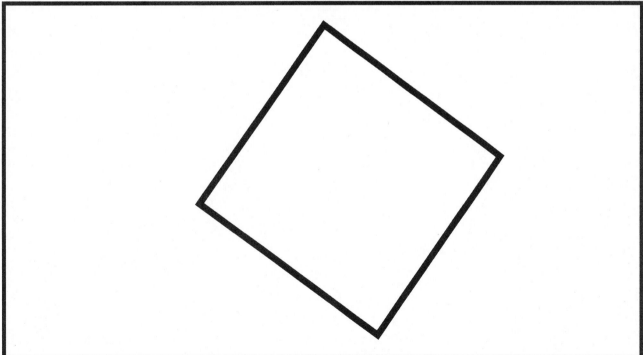

two-dimensional shape flashcards

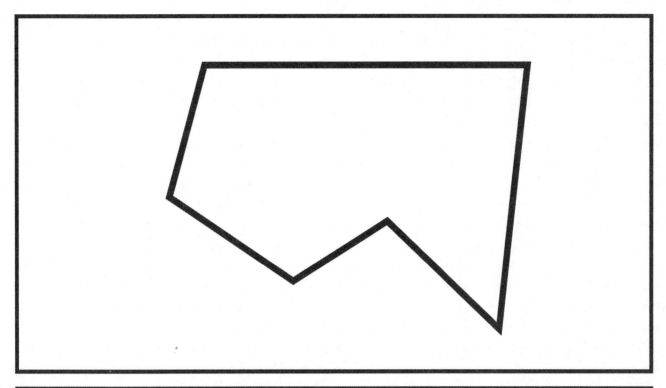

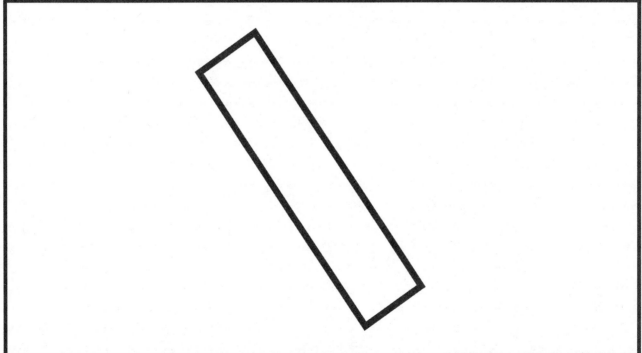

two-dimensional shape flashcards

Lesson 27:   Share and critique peer strategies for solving problems of varied types.

135

2-D SHAPES	3-D SHAPES
circle	sphere
triangle	cone
rectangle	cylinder
rhombus	rectangular prism
square	cube
trapezoid	
hexagon	

_____ corners	_____ corners
_____ square corners	_____ faces
_____ sides	_____ straight edges
Are all sides the same length?	Are all faces the same shape?
yes no	yes no

_____

shapes recording sheet

A

Number Correct: _____

Name _____   Date _____

*Write the number of dots. Try to find ways to group the dots to make counting easier!

1.	••		16.	••••• ••••
2.	•••		17.	••••• •••
3.	••••		18.	••••• •••••
4.	•••		19.	••••• ••
5.	•		20.	••••• •
6.	••••		21.	••••• ••••
7.	•••••		22.	••••• •••••
8.	••••		23.	•••• •••••
9.	••••• •		24.	••••• •••
10.	••••• ••		25.	••• •• •••••
11.	•••••		26.	••••• ••
12.	••••		27.	••• ••• •• •••
13.	••••• •		28.	•• ••• •• ••
14.	••••• •••		29.	•• ••• • ••
15.	••••• ••		30.	•• •• ••••

EUREKA MATH   **Lesson 28:**   Celebrate progress in fluency with adding and subtracting within 10   **139**
(and 20). Organize engaging summer practice.

©2015 Great Minds®. eureka-math.org

B

Number Correct: _____

Name _____   Date _____

*Write the number of dots. Try to find ways to group the dots to make counting easier!

1.	•		16.	••••• •••	
2.	••		17.	••••• ••••	
3.	•		18.	••••• ••	
4.	••••		19.	••••• •••	
5.	•••		20.	••••• •••••	
6.	•••••		21.	••••• ••••	
7.	••••		22.	••••• •••••	
8.	•••••		23.	• •••• •••••	
9.	••••• ••		24.	••••• •••••	
10.	••••• •		25.	•• •••••	
11.	••••• •••		26.	••• ••• •• ••	
12.	••••• •		27.	•• ••• ••• ••	
13.	•••••		28.	•• • •• ••	
14.	••••• ••		29.	•• •• ••	
15.	••••• •		30.	•• •• ••••	

Target Number:

Target Practice

Choose a *target number* between 6 and 10, and write it in the middle of the circle on the top of the page. Roll a die. Write the number rolled in the circle at the end of one of the arrows. Then, make a bull's-eye by writing the number needed to make your target in the other circle.

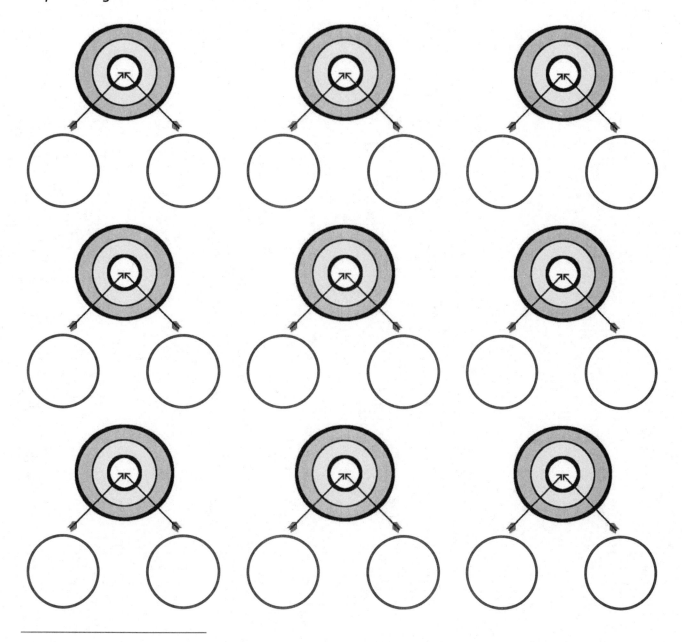

target practice

Lesson 28: Celebrate progress in fluency with adding and subtracting within 10 (and 20). Organize engaging summer practice.

143

Name _____   Date _____

 Race to the Top!

2	**3**	**4**	**5**	**6**	**7**	**8**	**9**	**10**	**11**	**12**

_____

race to the top

Lesson 28:   Celebrate progress in fluency with adding and subtracting within 10 (and 20). Organize engaging summer practice.

145

©2015 Great Minds®. eureka-math.org

Name _____   Date _____

Number Bond Dash!

Directions: Do as many as you can in 90 seconds.

Write the amount you finished here:

1.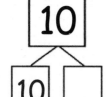
2.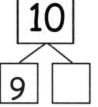
3.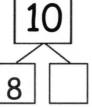
4.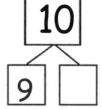
5.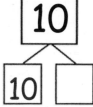

6.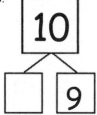
7.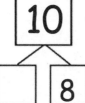
8.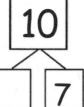
9.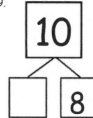
10.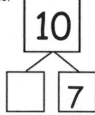

11.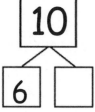
12.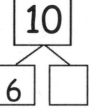
13.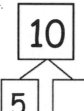
14.
15.

16.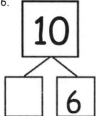
17.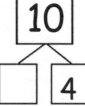
18.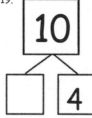
19.
20.

21.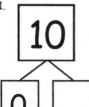
22.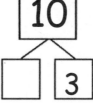
23.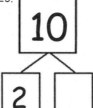
24.
25.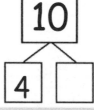

Credits

Great Minds® has made every effort to obtain permission for the reprinting of all copyrighted material. If any owner of copyrighted material is not acknowledged herein, please contact Great Minds for proper acknowledgment in all future editions and reprints of this module.